Unstuck and Unstoppable

The Unlikely Friendship With AI That Changed My Life

Unstuck and Unstoppable
Copyright © 2025 Abbie Boudreau
All rights reserved
ISBN: 979-8-9926472-0-4

No part of this book may be reproduced, distributed, or transmitted in any form or by any means, including photocopying, recording, or other electronic or mechanical methods, without the prior written permission of the publisher, except in the case of brief quotations embodied in critical reviews and certain other noncommercial uses permitted by copyright law.

Published by:
Seven Palms Publishing
Phoenix, AZ
www.abbieshop.com

Cover design by Abbie Boudreau
Interior design by Abbie Boudreau with AI assistance

This is a work of nonfiction. The information provided is based on personal experiences and research. The author and publisher make no representations or warranties with respect to the accuracy, applicability, or completeness of the contents. The advice given in this book is not a substitute for professional guidance.

For permissions or inquiries contact @AbbieBoudreau

Proudly printed in the United States of America

Dedication

To my parents and sisters—thank you for always listening, for believing in my ideas, and for trying to see my vision, even when I wasn't sure I could see it myself. Your support, especially during the last four years when I felt stuck, has meant everything.

To my children—who, even at their young ages, already understand that life is a series of ever-evolving chapters. Your curiosity, resilience, and belief in new beginnings inspire me daily to keep discovering what's next.

And to my husband—my steady ground through it all. You have been there through every shift, every uncertainty, and every reinvention. When I felt lost after my first career ended, you never doubted that I'd find my way again. You believed in me, pushed me to be my best, and reminded me that my dreams were far from over—they were just getting started. I love you.

Chapter 1

From Dream Job to Limbo Land

I used to know exactly who I was—until I didn't. It's been four years since I left my twenty-year career as a TV journalist. For many, being a network correspondent was the dream job. For me? Well, it's complicated. While it certainly started off that way—especially in the pre-kid chapter of my life—the last eight years of my career were more of a never-ending mental debate: Should I quit? Should I stick it out? Should I quit? (Rinse and repeat.)

When the pandemic hit, that debate finally ended—though not entirely by choice. I mean, *officially*, I quit, but let's be real: I don't know many people with an intact ego who'd accept an 80% pay cut. So, there I was, freshly "quit" and ready to figure out what was next. Except that's not what happened at all.

Instead, I was living in Los Angeles, suddenly thrust into homeschooling my two little kids, scrambling to remember second-grade math, and desperately trying to figure out who I was without a fancy job title—or a hair and makeup team.

As if my existential crisis wasn't enough, people kept asking:

- "Do you miss your old job?"
- "Are you going to go back to TV?"

The answers? No, and absolutely not.

Then came the inevitable follow-up: "So, what are you going to do now?"

I thought I'd have an answer. I thought it would be easy. Instead, I stirred, spiraled, and struggled with that very question.

Losing Myself

For two decades, I had been *someone*—someone with a career, a bio, a purpose. Losing that felt almost like a death.

My old life disappeared overnight. No more phone calls from frantic producers telling me to get on the next flight to cover a breaking news story. No more 2:30 a.m. alarms jolting me awake so I could go live on the West Coast on ABC News' *Good Morning America*. The life I had built—the one that had shaped me, defined me, consumed me—had vanished.

And there I was, standing in front of my bathroom mirror, staring at my own reflection like a stranger, asking myself: *Who am I now?*

I felt guilty that I couldn't just be happy. My job had drained me ever since I became a mother. I had been toying with leaving for nearly a decade. So why couldn't I embrace this new chapter? Why wasn't being a wife and mom *enough*? Why did I still crave *more*?

Maybe I had turned into the kind of person who would never truly be satisfied. Maybe I had lost the ability to be happy altogether.

What had happened to me?

I wasn't just questioning my next career move—I was realizing, for the first time, that somewhere along the way, I had lost *myself.* My entire identity had been wrapped up in being *the journalist.* Without that title, without the deadlines and the chaos and the urgency, I felt completely lost.

And because my job had been the reason we were living in Los Angeles in the first place, it wasn't long before my husband and I discussed the idea of moving back to Arizona, where we had first met 15 years earlier. Six months later, we packed up our belongings and left.

It was all too much. In half a year, I lost my career, my home, my closest friends, my California dream, and, most painfully, my entire sense of identity. On the surface, it might have looked like a fresh start. I wanted it to be. But it wasn't. It was too much change, too fast, and it left me completely unsure of who I was or what the heck I was supposed to do next.

The Stuck Years

For four years, I cycled through endless career ideas:

- Realtor? Author? Stay-at-home mom?
- Help my husband with his mortgage business? Website developer?
- Become a teacher? Or maybe a social media influencer? (I told you I was spiraling!)

A few of these ideas even made it to the "trying" phase. I launched three websites, a few Instagram pages, started three manuscripts that never saw the light of day, and brainstormed a hundred other ideas that felt exciting for about five minutes before fizzling out.

During those four long years, I convinced myself that I had become indecisive—or worse, *uninspired*. None of my ideas seemed worth the effort, so I stayed *stuck* in this weird limbo: a kind of analysis paralysis, a start-and-stop, why-even-bother mentality. And the hamster wheel just kept spinning.

Enter ChatGPT (Yes, Really)

Fast forward to three days ago. I was having yet another career crisis (this time, deciding if I should become a general contractor) when I turned to my latest sounding board and new BFF: ChatGPT. Here's a portion of my back and forth.

MY QUESTION: *"I'm starting to think again about getting my GC license in Arizona and I want to find out the easiest license to get me started. Can you help me with this?"*

CHAT: It gave me a very long answer.

MY RESPONSE: *"It just seems like that would take a long time. I'd be almost 50 years old by the time I got my license. What's another job idea that I could do that has less requirements to getting started?"*

CHAT: Another very long and detailed answer.

MY RESPONSE: *"I don't like any of those ideas. Maybe event planning but even that does not sound fun."*

At this point, even Chat was probably rolling its digital eyes at me.

MY RESPONSE: *"I appreciate those ideas, but still, I am not feeling like those are worth pursuing. I feel like *there's always a reason not to do something. I just need to do something and get it started. I just cannot decide what to do."*

CHAT: *"It sounds like you're at a crossroads and feeling the weight of indecision. The truth is, there's rarely a "perfect" choice—what matters most is picking something that feels **good enough to start**, even if it's not perfect. The clarity and fulfillment often come after you take action, not before."*

MY RESPONSE: *"Maybe I should write a book about what you just said. I am sure a lot of people get **stuck** at this point."*

Just like that, I had an epiphany: I wasn't indecisive, uninspired, or lazy. I was *stuck*. It was the perfect word to describe my mental state—so simple, yet so profound. For years, I had tried labeling myself with a million other words, but none of them captured what I was truly feeling. When I finally wrote the word *"stuck"* and equated it with myself, it was like a lightbulb went off. I finally understood what was going on—and, even more importantly, what I was going to do about it.

What This Book Is About

Unstuck and Unstoppable is a deep dive into what it really means to feel *stuck*—using ChatGPT as a thinking partner to explore how we got here, why moving forward feels so hard, and what it actually takes to break free and feel fully alive again. It's about giving ourselves permission to dream, not like we used to, but in a way that reflects who we are now. Because our lives have changed. Our dreams have changed, and we owe it to ourselves to stop dismissing them and start chasing what truly lights us up. This book is about finding your entrepreneurial spark again and building something that feels real and good—whether it's bold and ambitious or quiet and personal or somewhere beautifully in between.

This isn't a guide to building a billion-dollar empire or becoming the next big thing. It's for the stay-at-home moms wondering if they're more than "Mom." It's for people who left a career and don't want their best stories to live in the past. It's for anyone who's tired of feeling stuck and wants to move forward—not perfectly, not with a grand plan, but with curiosity, courage, and a monthly

ChatGPT subscription that I now consider an essential business expense.

This book is for you. And it's for me too. (Now, let's see if I can follow my own advice and *actually finish* this thing.)

Chapter 2

Famous Pivots and the Art of Not Staying Stuck

Before we dive into how AI can help reshape your next chapter, let's zoom out for a moment. Long before the dawn of ChatGPT, people were reinventing themselves the hard way—armed with nothing but grit, self-belief, and a willingness to leap without a blueprint for success. If they could pivot without powerful tools like the ones we now have at our fingertips, just imagine what's possible for us.

It's easy to assume that people with household names—actors, athletes, and entrepreneurs—had it all figured out from the start. That they were born with a roadmap, marched confidently toward success, and never once spiraled in bed at 2 a.m. wondering what they were doing with their lives.

Yeah, no.

The truth is, even the most successful, accomplished people have hit walls. They've doubted themselves. They've realized, often painfully, that the path they were on wasn't working anymore. And here's what is most interesting: most of them didn't have a flawless plan for what came next. They pivoted anyway.

It wasn't until I truly understood that I was *stuck* that my perspective began to shift. Once I saw my own struggle clearly, I started noticing other people who had made major pivots in their lives. Who knows if they were stuck for years like me, but I imagine they shared similar fears and insecurities along the way.

Let me be clear: This isn't a list of people I admire because they walked red carpets or won awards. These are people who *chose* to move forward when they could've stayed stuck. *People who faced doubt and did it anyway because they refused to believe that one chapter defines their whole story.*

So, if you're sitting there wondering if it's too late to switch gears, I hope these stories inspire you and remind you: *It's not about having it all figured out. It's about being brave enough to try.*

Julia Child: From Government Spy to Gourmet Icon

At thirty-six years old, Julia Child didn't know how to cook. Not even a little. Before becoming the world's most beloved chef, she worked for the U.S. government as a research assistant during World War II, basically a desk job. It wasn't until she moved to France in her late 30s and tasted a life-changing meal (*boeuf bourguignon, naturally*) that she thought, "*What if?*"

What followed was years of hard work, cooking school, and doubt. Julia didn't publish her first cookbook until she was forty-nine, and her iconic TV show, *The French Chef,* didn't debut until she was in her 50s. Imagine if she'd said, "I'm too old to start something new," or, "What if people don't take me seriously as a chef?" Her story is proof that age and doubt are no match for passion and persistence—and the culinary world is far richer for it.

Vera Wang: From Figure Skating Dreams to Bridal Couture

If Vera Wang had followed her original plan, you wouldn't know her name—you'd just see her figure skating highlights from the '60s. Vera's dream of becoming an Olympic skater didn't work out, so she pivoted to fashion. But here's the twist: her *second* plan didn't kick in until her 40s.

After a long career as a fashion editor, Vera hit a new crossroads: she couldn't find a wedding dress she loved when she got married at age forty. Instead of settling, she thought, *"What if I just design one myself?"* That "hobby" turned into a multi-billion-dollar bridal empire. Her decision to take a chance redefined bridal fashion forever, proving that bold ideas can reshape entire industries.

Colonel Sanders: KFC After 60

Picture this: You're sixty-five years old, unemployed, and living off social security. What do you do? If you're Colonel Harland Sanders, you start perfecting your fried chicken recipe and driving around the country trying to sell it.

Before the white suit and the Kentucky Fried empire, Sanders held more jobs than most of us could dream up: farmer, streetcar conductor, insurance salesman, and even a gas station operator. It wasn't until his mid-60s that he found his true calling.

The moral? If your first (or second, or fifth) plan doesn't work out, keep going. Even if your success comes with wrinkles, it's still success.

Harrison Ford: From Carpenter to Star Wars

Before he was Han Solo, Harrison Ford was a carpenter—literally building tables and cabinets to pay the bills while taking acting gigs on the side. At one point, Ford famously built doors for *Star Wars* director George Lucas, who later gave him the role that changed his life.

Imagine if Ford had thrown in the towel after a few bad auditions and said, "Maybe carpentry is it for me." Instead, he kept one foot in what paid the bills and the other in his dreams. That's the thing about second chapters—they don't always start with a dramatic exit.

Sometimes, they sneak up on you while you're sanding a cabinet door.

Toni Morrison: From Editor to Award-Winning Author

Toni Morrison spent years editing other people's books before publishing her own. She didn't write her first novel, *The Bluest Eye*, until she was thirty-nine. At forty-eight, she published *Song of Solomon*, which put her on the literary map. And at fifty-six, she won the Pulitzer Prize.

It would have been easy for Morrison to stay where she was—successful, comfortable, behind the scenes. But she didn't. She chose her second chapter, and in doing so, gave the world some of its most powerful and enduring stories.

The Realization (And a Greenhouse Dream)

Here's what all these stories have in common: uncertainty. Every one of these people had to take a deep breath, quiet the voice that said, "*You can't do that,*" and step into something new.

Did they know for sure it would work? Nope. Did they keep going anyway? Absolutely.

For me, I think the answer has been sitting quietly in the background all along, waiting for me to notice it.

The truth is, I know what I love. I love painting, embroidery, and needlework. I dream of building a beautiful greenhouse and living garden-to-table. But here's the thing, the old me would have thought: None of those ideas felt *big enough*. Important enough. "*Career*" enough. I'd tell myself, "Those are hobbies, not careers."

And I'd think, "Am I really going to go from high-profile journalist to greenhouse goddess with an embroidery hoop in hand?" The doubt was loud, persistent, and totally paralyzing. So I would push those ideas away, along with my passion. I'd bury them, ignore them, convince myself there had to be something bigger. Something more *worthy* of my past.

Maybe I need to start thinking more like Vera and Julia. Who says I can't make a career shift without a clear plan? Maybe my greenhouse project isn't just a wistful daydream—it could become a book, a YouTube channel, or a way to inspire others to embrace garden-to-table living in the desert. And maybe those quiet hours I spend painting and embroidering aren't just hobbies after all. Maybe they're the beginning of something real—something I can share, sell, and build a new kind of life around.

Saying these dreams out loud *for the first time* is like flipping on a light switch in a dark room. For so long, I've kept these ideas locked up, convincing myself they weren't "big enough" or "important enough." But now, I'm daring to imagine what they could become. Putting these ideas on paper *for the first time*, these aren't just passing thoughts anymore; they're the beginnings of something real.

What's my pivot? Without overthinking it to death, maybe I've just answered the question that's been hanging over me for years. And here's the most surprising part: When you realize the problem wasn't that you were incapable—but just *stuck*—finding clarity isn't nearly as hard as you thought.

Your Turn: Flip the Light Switch

Take a moment to think about those quiet ideas you've been keeping locked up. You know, the ones that feel "too small" or "not important enough."

Step 1: What's a dream you've been too afraid to say out loud? Write it down, no matter how big or small.

Step 2: What's stopping you from pursuing it? List every excuse, fear, or doubt holding you back (yes, even the silly ones).

Step 3: Now, ask yourself: What if I ignored the doubt for just one minute? What could this dream look like if you stopped overthinking and let yourself imagine the possibilities?

Chapter 3

What's Next? (AKA, The Question That Might Break You)

There's nothing quite as humbling as standing at the crossroads of your own life, staring at that relentless, flashing question: *What do I do next?* It doesn't matter whether you had a high-powered career with a fancy title or one of those quietly impactful, unsung hero jobs. When it comes to creating a second act for yourself, it's probably harder than you'd expect. But there's a reason for that: You've changed.

Why 'Small' Feels Wrong

We live in a world that glorifies "big"—big careers, big milestones, big dreams. So when your next step feels quiet, personal, or dare I say, *small*, it's easy to dismiss it. We think, "*How could something so simple really matter?*" But sometimes, small is exactly what you need. The problem is that comparison creeps in.

It doesn't matter if you're comparing yourself to your old self, your neighbor, or that influencer you follow online—comparison will always make you feel like you're falling short. Somewhere along the way, we've been taught that "big" is the only thing worthwhile. But let me tell you something: *the size of your dream doesn't determine its value.* What matters is that it means something to *you*—even if no one claps for you but yourself.

The Woman Who Was (Annoyingly) Right

When I was interviewing for a job as a network news correspondent, the woman in charge of hiring talent asked me what I'm pretty sure was a blatant HR violation: "Do you want to be a mom someday?" Without skipping a beat, I told her yes—and added that I believed I could have it all. Her response? A curt, "You can have it all, but *not all at the same time.*"

I was furious. How dare she suggest that I couldn't juggle motherhood and a top-tier journalism career? It didn't matter to me that she was twice my age, had kids of her own, and had probably already tried—and failed—at *having it all.* I convinced myself that maybe *she* couldn't, but I was young, determined, and ready to prove her wrong. For years, I carried that stubbornness with me, juggling it all—the demanding career, the accolades, and eventually, *the family.*

That's when something started to shift. The effort it took to keep all the plates spinning left me exhausted. I was constantly giving 100% to everything, but I still felt like I was falling short.

It wasn't until much later that the realization finally hit me: "having it all" is a myth sold to women who are too young to know any better. And despite being warned, I couldn't help but desperately chase that myth until I couldn't any longer.

Because the truth is, you can't give 100% to your job and 100% to your family at the same time. Math doesn't work that way. And even if it did, who would have anything left for themselves?

I finally saw it clearly: I didn't want to have it all anymore. The relentless pace of my career left me exhausted, stretched thin, and missing moments I knew I could never get back. I didn't want to piece together memories of my kids' childhood from rushed mornings and late-night check-ins. I wanted to be present, to really experience it with them. But that choice came with a realization—one I'm still unpacking as I write this: my dreams look different now, and maybe...that's okay.

The truth is, I've changed. I can't approach my next chapter the way I approached my first because my life—and what matters to me—looks completely different now. *The key to moving forward is accepting that change. Until we do, we'll stay stuck, trying to chase dreams that no longer fit who we are.*

The Silver Lining (Yes, There Is One!)

The good news is, this isn't about starting from scratch. Even if it feels like your old career is a distant memory, the skills you've built along the way are still with you—and they're surprisingly transferable.

For example:

- If you were a journalist (like me), you're great at storytelling, meeting deadlines, and asking tough questions.
- If you were a teacher, you can manage chaos and inspire people (aka future clients).
- If you were a stay-at-home parent, congratulations—you've got CEO-level problem-solving skills and the patience of a saint.

Skills You Forgot You Had (Until Now)

Think about the skills, strengths, and superpowers you gained from your previous chapter. You're more equipped than you think, and this list is proof.

Chapter 4

It's Not Me, It's *You*.

Here's the thing they don't tell you about being stuck: it's not just exhausting—it's infuriating. Not only are you mad at yourself for being indecisive, but now you're also mad at everyone else for not magically solving your life for you.

And trust me, I've been there. I've stared my husband dead in the eyes and asked, "Why won't you just tell me what to do?" as if he's some sort of career GPS. To which he says, with his usual supportive tone, "I think you'd be amazing at anything." Cue my meltdown.

The Spouse (or Best Friend) Effect

Let's talk about the poor, innocent cheerleaders in your life. Maybe it's your spouse, best friend, or mom—someone who genuinely wants the best for you but has grown accustomed to you running countless ideas past them with wild enthusiasm, only to pivot to a totally different idea days or weeks later. They've learned to tread carefully, so instead of calling you out for your inability to make up your mind, they say things like:

- "You'd excel at whatever you put your mind to!"
- "I think all your ideas are great!"

And while that sounds supportive, what you hear is:

- "Why won't you just pick already?"
- "I'm tired of hearing about this."
- "Can you go figure this out quietly somewhere else?"

Lost in the Fog

Anger and desperation can make you do some truly questionable things. *When you're stuck, the longing for purpose—for any sense of forward momentum—can cloud your judgment.* It's that gnawing urge to just do something, anything, to feel like you're moving forward. And in that haze, you might grab onto the first thing that comes your way, even if it's a bad idea. That's exactly how I found myself agreeing to go into business with someone I barely knew.

At first, it felt exciting. Finally, I had direction. A plan. But the excitement faded quickly when reality set in. I was doing all the work—leveraging my journalism background to write every blog post, Instagram caption, and article—while my so-called business partner spent her days going out to lunch with friends and playing pickleball.

As the imbalance grew, so did my anger: "Why did I let this happen? Why didn't I see the red flags? Why am I stuck doing everything?" The answer, I realized, was *desperation*. I was so eager to feel *unstuck* that I ignored my gut and threw myself into a situation that wasn't right for me.

That's the thing about desperation: it makes you vulnerable. When you're so focused on escaping the pain of being stuck, you stop looking at where you're going. And when things inevitably fall apart, the anger that follows isn't just aimed at the situation—it's aimed at yourself. That self-directed anger comes from realizing you knew better but did it anyway. It's a brutal kind of clarity.

In my case, it was a comment from my kids that made everything click. They asked, "Mom, why is she (my former business partner) going on vacation all of the time while you're working all day?" Then came the zinger: "It's like you're her Siri. You're just there to answer all her questions." Ouch. But also—accurate.

That unfiltered observation cut through my rationalizations and showed me what I'd been too afraid to admit: *I was letting myself be used because I was so desperate to feel useful.* Those words made me stop and reflect on how my vulnerability had led me straight into a bad decision.

Looking back, it's no wonder I was so angry—at the situation, at my "business partner," and mostly at myself. Anger often shows up when we realize how we've ignored our instincts or allowed desperation to steer us off course. But that anger can also be a gift, a signal to step back, reassess, and do better moving forward.

The Spiral of Misplaced Expectation and Blame

If you're anything like me, you've probably found yourself in one (or all) of these situations:

- **Getting mad at someone for not giving you an answer.**
  "Why won't you just tell me what I should do? You know me best!"
- **Getting mad at someone for liking all your ideas.**
  "You can't possibly think *all* of these are good. Stop cheerleading and start guiding!"
- **Getting mad at someone for being neutral.**
  "Why don't you care more about my existential crisis?"
- **Getting mad at someone for having an answer you didn't like.**
  "Okay, that's a terrible idea. Are you even listening to me?"

Why This Happens

This spiral of misplaced frustration and anger doesn't solve anything. Here's the brutal truth: Being mad at other people is easier than being mad at yourself. And when you're stuck, you're already beating yourself up enough, so the frustration spills over.

But guess what? It's not their job to figure this out for you. And even if they tried, you'd probably reject their advice anyway. (You know it's true.)

Prompts for Progress

Step 1: Think about the person you've unfairly snapped at during your "What should I do next?" crisis. Write their name below. (Don't worry, you're not showing them this.)

Step 2: Next to it, write down the nicest thing they've said to you during this whole ordeal. (Yes, even if it was annoying at the time. **And this time,** *try believing what they said.*)

Chapter 5

The Truth About Staying Stuck

Let's call it what it is: we're *stuck*. Whether it's been weeks, months, or years, here we are—knee-deep in mental quicksand.

We're overthinking. We're second-guessing. We're so tangled up in "what ifs" that we've practically wallpapered our brains with them. "What if it's too late?" "What if I'm no good?" "What if I pick the wrong thing and waste my time and money?"

And if we've had a successful career before? The pressure is even heavier. Because now it's not just about what comes next—it's about living up to everything we've already done. It's like trying to write a sequel to our own lives while a voice in our heads whispers, "Eh, the first one was better."

What Is the Stuck Zone?

Before we can break free—with a little help from AI—we need to understand what the Stuck Zone really is, how it sneaks into our lives, and how we got here in the first place.

The Stuck Zone is that sticky, frustrating place where every decision feels enormous, and we're paralyzed by too many possibilities—or worse, by the fear of making the wrong choice.

- It's the "I want to do something meaningful, but... what if it sucks?" phase.
- It's the "Maybe I should go back to school... No, I'm way too old" spiral.
- It's the endless loop of internal debates, followed by no meaningful action.

What makes the Stuck Zone so disorienting is that our minds are stuck in the past. We're still thinking and dreaming the way we used to—back when our lives, identities, and priorities were completely different. *But we've changed.* Our lives have changed. *And yet, we keep reaching for dreams that no longer fit who we are now.* Of course they feel impossible. Of course they fall flat. Deep down, we know those old dreams have run their course. But when we're swirling in self-doubt, it's hard to see that clearly. Instead, we ask, "What's wrong with me?" "Why doesn't anything *feel right* anymore?"

The truth is: there's nothing wrong with us. *We're just trying to force a future that doesn't match the people we've become.*

The Life That Came Before

Second chapters are filled with second-guessing.

When I was starting out in my career, I was single, child-free, and laser-focused on proving myself. I poured every ounce of energy into becoming successful. Back then, I could ask, "What do I want to do with my life?"—and answer it with bold, ambitious clarity.

But life doesn't stay static. I've changed. I've built a family-centered life I love. And with that shift, my priorities have changed too. I no longer have the same kind of time, energy, or desire for a career that consumes everything. I used to chase big dreams with tunnel vision. Now, I want something meaningful that still leaves space for the life I've built.

Take one of my old dreams—starting a video production company and creating hard-hitting investigative documentaries. It sounds amazing. Impactful. Purpose-driven. But it doesn't fit with the version of my life I'm living now. Trying to force that old dream into my current reality only leads to more frustration—and more feeling stuck.

Permission to Dream Differently

Here's what we often forget: our dreams aren't what they used to be. *And they're not supposed to be.*

In the quiet moments, when I'm really honest with myself, a new dream starts to whisper. What if I could build a slower, more creative life—filled with handmade art, a greenhouse, and maybe even teaching others about garden-to-table living? That dream feels tender. Personal.

But when we're stuck, dreams like that don't always compute. They feel too different. Too small. But maybe the problem isn't the dream—it's the outdated lens we're using to judge it.

So, sure, staying stuck might feel safe. But long-term? It's the riskiest choice of all.

The Day I Learned to Burn the Shelf Down

Before I stepped into the world of second chapters (and third and fourth chapters, if we're being honest), I spent 20 years as a journalist—first as an investigative reporter and later as an entertainment correspondent for ABC's *Good Morning America*. I covered everything from hard-hitting stories to interviews with Hollywood's biggest names.

That's how I ended up sitting across from Erika Jayne, the unapologetic *Real Housewives of Beverly Hills* star, who at the time was a contestant on *Dancing with the Stars*, a hit show on ABC, which meant I was assigned to interview her.

Here's the thing: I've interviewed tons of fascinating people. Most of the time, I'd walk away with something *interesting*—a fun fact, a quirky anecdote—but I wasn't always *inspired*. I certainly wasn't expecting a Maya

Angelou moment from Erika Jayne, but that's exactly what I got.

When I asked her about reinventing herself as a pop singer at forty-five, she looked me straight in the eye and said, "No one puts a shelf life on me."

I loved that comment. She didn't just *say* it. She *meant* it.

She wasn't asking for permission to do something bold and unexpected, she was *declaring* it. I often think about that one line. Who says your age, your past career, or anyone else's expectations get to decide what you do next?

And sure, Erika's reinvention was big and bold—front and center on reality TV. Mine might look more like wildflowers in a greenhouse or art stitched slowly by hand. But here's the beautiful part: the *size* of the dream doesn't matter. *The courage to change is what unites us.* Reinvention doesn't have to be loud or public. It just has to be honest. Whether you're launching a business, planting a garden, or finally doing that thing you've talked yourself out of for years—the shelf life doesn't exist unless you put it there.

Excuses, Exposed: Time to Rewrite the Script

Let's flip the script. Right now.

Step 1: Write down the worst excuse you've been telling yourself. The dumbest one. Don't hold back. "I'm too old." "I don't have the time." "What's the point?"

Step 2: Rewrite it like you're coaching a friend.

- "I'm too old" becomes "Age is the least interesting thing about me."
- "I don't have time" becomes "I can start with 10 minutes a day."
- "It's been done before" becomes "Yeah, but it hasn't been done by me."

Chapter 6

Meet Your Stuck Personality (Spoiler: I'm the Perfectionist)

Let's get one thing straight: feeling stuck isn't a personal failure—it's practically a personality trait. I know this firsthand.

I've spent more time wrestling with indecision about my next chapter than I care to admit. And it's not because I can't make big decisions—I make them all the time. I designed my entire home remodel, picking out every detail. I decided to homeschool my daughter when she needed extra help. I make big decisions all the time and I handle them confidently. So, you'd think figuring out my next chapter would be a breeze. But it's not. Why? Because I'm a perfectionist.

The same personality trait that helped me achieve great success in my previous career as a network journalist, was keeping me stuck this time around. I hadn't realized how much this was holding me back. I chalked it up to indecision or maybe a lack of time and energy. But once I realized that my obsession with making the "perfect" decision was actually what was keeping me stuck, the pattern finally started to make sense.

And here's what I've learned: it's not just me. Most of us have a "stuck" personality type that shapes how we approach decisions—or avoid them. Maybe you're an overthinker, caught in an endless loop of pros and cons. Maybe you're a people-pleaser, too worried about what others will think. Or maybe you're a safety seeker, reluctant to leave your comfort zone.

The good news? Once you know your stuck personality, you can start working *with* it instead of against it.

The Four Stuck Personalities (Pick Yours)

The Overthinker

How You Got Stuck: Your brain is like a hamster wheel on steroids. You analyze every option, consider every outcome, and still end up...nowhere. Because what if you pick wrong?

How to Get Unstuck:

- **Try the 80% Rule:** Overthinkers often get stuck searching for 100% certainty – but, that almost never happens. If you're 80% sure about a decision, that's probably enough to move forward. Think of it like online shopping. You read the reviews, compare options, and maybe even leave the item in your cart for days. But at a certain point, you have to make a decision to buy or not to buy. If you were *mostly* sure it was the right choice, you'd probably go for it. So why not apply the same logic here? If you're 80% sure the idea is worth pursuing, maybe that's enough to take the next step. Trust that you can figure out the rest along the way.
- **Reverse the Pros and Cons:** Instead of listing the pros and cons of each option, write down the worst-case scenarios if you *don't* decide. (Staying stuck is usually the scariest outcome.)

- **Try a Coin Flip Trick:** Assign each option to a side of a coin, flip it, and then ask yourself: "Am I happy with the result?" If not, you know your gut is pointing you in the other direction.

The Perfectionist (Hi, That's Me)

How You Got Stuck: If it's not perfect, it's not worth doing. You spend so much time crafting the "right" plan that you never actually start.

Let me tell you, being a perfectionist is exhausting. For example, I dream about creating a garden-to-table greenhouse in Arizona or figuring out a way to sell my artwork and paintings. But every time I start to imagine those ideas, my brain immediately jumps to: *What if the greenhouse isn't Instagram-worthy? What if no one buys my embroidery? What if I spend months on this and it all flops?*

That's the thing about perfectionism—it turns every step into a high-stakes decision, which makes starting feel impossible.

But here's the truth I've had to learn (and relearn): Perfect doesn't exist. Waiting for the perfect time, the perfect plan, or the perfect outcome means you'll be waiting forever.

How to Get Unstuck:

- **Pick One Small, Imperfect Step:** For me, that might mean planting one vegetable in a pot instead of building the whole greenhouse. Or post one embroidery design on Etsy before creating a full shop. The point is to start small and accept that it doesn't have to be life altering to be meaningful.
- **Set a "Good Enough" Goal:** Instead of imagining a fully functioning greenhouse, I could aim for a single herb garden to start. As for my art, I could set a goal of uploading just three items, even if they aren't professionally photographed or perfectly described.
- **Create a "Messy Challenge":** Commit to doing something badly on purpose. Maybe that could mean posting an embroidery piece that feels a little unfinished or trying to grow a plant I'm not sure will

thrive. The idea is to prove that starting imperfectly won't end the world—it'll just get the ball rolling.

- **Ask Yourself: What's the Worst That Can Happen?** Honestly, what's the worst that could happen if the greenhouse takes longer than planned, or the Etsy shop only gets one sale? Nothing catastrophic. And even if it doesn't go as planned, at least I'll learn something along the way.

The People Pleaser

How You Got Stuck: You're so busy worrying about what others think that you forget to ask yourself what *you* want. (Shoot! Double whammy for me, I guess. I have this personality trait too. No wonder I've been stuck!) Every decision becomes about meeting someone else's expectations—your family's, your friend's, or even a random acquaintance's—and your own desires get buried under a pile of "What will they say?" It's like living your life as a perpetual approval-seeker instead of the main character.

How to Get Unstuck:

- **Ask the "No One Knows" Question:** Ask yourself, "If no one ever found out about this decision, would I still want to do it?" If the answer is yes, move forward with confidence.
- **Create a "Me First" Hour:** Block out a couple of hours a week to focus solely on what *you* want to explore or try—no outside input allowed. Treat it like a sacred appointment with yourself.
- **Reclaim Your Spotlight (One Tiny Rebellion at a Time):** Pick something you love that no one else in your life really "gets." Now do it anyway. Crochet a neon sweater. Order the weird menu item. The goal isn't rebellion for rebellion's sake—it's reminding yourself that you're allowed to do things just because they make *you* happy. Tiny rebellions build big confidence.

The Safety Seeker

How You Got Stuck: Your comfort zone is your happy place. The unknown feels like freefalling without a parachute, so you avoid it at all costs. You convince yourself that staying put is the "smart" choice, even as your dreams gather dust on a shelf marked "Someday." Instead of looking forward, you spend a lot of time gazing into the past, clinging to previous successes and accomplishments as proof of your capability. You know you could do whatever you set your mind to, but the thought of leaving your safety zone feels too risky. So, you rest on your laurels and *convince yourself that you've already done enough.* Deep down, though, you know that stepping out and going for it would make you feel infinitely more alive and fulfilled.

How to Get Unstuck:

- **Take a "Mini Risk":** Do something tiny but slightly out of your comfort zone. Send a casual email to someone in a field you're curious about, or take one class that interests you.
- **Turn the Unknown Into the Known:** Do deep research on one small aspect of your next step. For example, if you're considering a new career, schedule an informational interview or read a book

about it. Knowledge makes the unknown less intimidating.

- **Track Your Wins:** Create a "Courage Journal" where you jot down every small risk you take and the result. Seeing your wins—even tiny ones—builds momentum and confidence.

Why This Chapter Matters

Understanding your stuck personality isn't about changing who you are—it's about recognizing how you operate and finding ways to work through it. Whether you're an overthinker, a perfectionist, a people-pleaser, or a safety seeker, the key is to stop letting your tendencies control you and start using them to your advantage.

For me, that's meant letting go of perfect, picking small steps, and laughing at my own ridiculous standards. (And yes, I'm still working on it. *Progress, not perfection, right?*)

Get to Know Your Stuck Personality

Identify your stuck personality and how it's been keeping you stuck. (And remember, if you are struggling to figure out which personality type you are, it's no big deal. You can always come back to this chapter later, after you gain more insight into your life and personality traits from ChatGPT. You will be shocked at how well AI can assist you with getting to know yourself!)

Chapter 7

The Toe Dip Tango

Let's talk about toe-dipping. You know the drill—testing out an idea just enough to say you *tried*, but not enough to actually commit. It usually starts with a spark of excitement, maybe even a Canva account, and ends with a quiet "Eh, maybe this isn't for me." When we don't have a clear direction or dream we're fully invested in, toe-dipping becomes our default. We end up saying yes to whatever floats by—kind of a "Sure, why not?" mentality. And if you've been there, trust me, you're in good company. I might even be the president of the club.

But here's what I am beginning to realize: toe-dipping isn't always as harmless as it sounds. Sure, it feels safe—it lets us claim we're trying without really putting ourselves out there. But after enough half-hearted attempts, toe-dipping can leave us feeling *more stuck*, not less. And worse? It keeps us from pursuing the things we're truly passionate about, the things that might actually wake us up.

My Greatest Hits of Toe-Dipping

In the past four years, I have:

- Launched three websites. (None of which I remember the passwords to.)
- Started multiple Instagram accounts that I have abandoned.
- Helped a general contractor with interior design. (Which sounds fancy until you realize your clients are also stuck in a cycle of *design* indecision.)
- Offered to help my husband with his mortgage company. (Despite knowing that spreadsheets make me want to cry.)
- Attempted to homeschool my daughter. (And then promptly hired a teacher when we both agreed it wasn't working.)

Each of these started with big hopes and Pinterest boards... and ended with me quietly walking away, muttering, "Nevermind."

Why We Toe-Dip

Toe-dipping offers the illusion of progress—it allows us to test the waters just enough to say, "I gave it a shot," without the weight of full commitment. But it also means we never really find out if something could've worked because we're too busy moving on to the next shiny idea.

Here's the thing, though: *so many of the ideas we try aren't even things we're passionate about.* They're ideas that sound good on paper or might impress our friends and family—"Oh, she's starting a business? How exciting!"—but they don't actually fill us up. And deep down, we already know they won't work. Without passion, we don't have the energy to stay engaged, especially when life (and helping our kids with their homework at 10 p.m.) inevitably pulls us in another direction.

Honestly, most of these toe-dip ideas weren't for me—they were for the people around me. They were to prove I was still doing *something*. But you know what I've realized? The only person judging me this harshly is myself. It's not my family or friends who are whispering, "Wow, she used to have it all, and now look at her." That's all me. I'm the one holding myself back, stuck in a cloud of self-doubt, instead of embracing ideas that actually bring me joy.

What If We Took a Different Approach?

As I've been writing this book, I've started to see my pattern more clearly. What if, instead of dismissing my smaller, more personal dreams—like figuring out a way to sell my art or building the greenhouse I've always imagined—I took real steps toward them? What if, instead of tiptoeing into every idea just to say I tried, I allowed myself to dive into something that actually excites me?

The truth is, I'm starting to love the thought of telling my family and friends, "I'm going to create and sell my embroidery art." A year ago, I would've obsessed over what they might say behind closed doors. "Is she serious? Is this some kind of mid-life crisis?" But now? Screw it. I'm tired of living in fear of what others might think. *It's time to stop letting imaginary critics (and my perfectionist, people-pleasing personality) keep me stuck.*

Here's the shift I'm making: Instead of chasing ideas I know I'll abandon at the drop of a hat, I'm going to pursue what truly matters to me—without measuring my new goals against the life I used to live. No more toe-dipping. It's time to cannonball. It's time to commit to a new passion—without a perfect plan, and without worrying about what anyone else might think. It's time to discover that thing I'm meant to do—the thing that lights me up and makes me say, *"Oh, so this is what it feels like to be alive again."*

Can Passion Pay the Bills?

There's a growing body of research that suggests loving what you do doesn't just make life better—it can actually lead to long-term success. A study from the University of Warwick in England found that happier workers are 12% more productive than their less-happy counterparts. And it's not just about efficiency. When people love what they do, they tend to stick with it longer, building skills and opening doors to opportunities they couldn't have imagined at the start.

So, what if the things we label as "hobbies"—those activities we could lose ourselves in for hours—are actually the seeds of something more? *Studies show that passion fuels perseverance, and perseverance can lead to success.* It's not about launching a multi-million-dollar business tomorrow. It's about starting small, finding happiness in the process, and *trusting that what you love has value.*

Who's to say that stitching flowers onto a hoop or growing a backyard full of veggies can't lead to something bigger? The idea of turning what lights me up into something sustainable has taught me this: *passion isn't just a starting point—it's the foundation for building a life that feels fulfilling.* And if that's true, the possibilities are endless.

Chapter 8

Finding Your Big Idea with a Little A.I. Magic

Up until now, we've explored what it truly *feels* like to be stuck—the doubts, the overthinking, the reasons so many of us struggle to move forward. We've even started the process of asking key questions to uncover new passions, those little entrepreneurial sparks that just *might* have the potential to grow into a second career. But I don't just want to *talk* about being stuck—I want to show you exactly how I started pulling myself out of it. And, somewhat unexpectedly, a huge part of that came from using AI.

I know, it might not be the first tool you think of when it comes to reinventing your career but hear me out: AI didn't magically hand me my next chapter on a silver platter, but it *did* become the best brainstorming partner I never knew I needed. It helped me challenge my own thinking, push past indecision, and, ultimately, start moving forward. So, as we shift into the next section of this book, I want to walk you through exactly how I used AI to break free from the *stuck zone*—and how you can, too.

AI: Myth, Mystery, and Your Secret Weapon for Getting Unstuck

Before we dive into *how* to use ChatGPT to brainstorm ideas, let's talk about the elephant in the room—AI has a bit of a reputation. Some people fear it, some think it's cheating, and others assume it's just a glorified search engine. And, of course, there's the whole *robots-taking-over-the-world* narrative.

I get it. AI can feel intimidating, mysterious, even a little unsettling. But here's the thing: it's not going anywhere. And while we can debate its place in the world, one thing is certain—it can be an incredibly powerful tool to help you get unstuck. Used the right way, AI won't replace your creativity or your ideas; it will help you uncover them. It can quiet the overthinking, break the analysis paralysis, and—bonus—your spouse, best friend, or go-to sounding board will probably thank you. Instead of circling the same old conversations, now they get to cheer you on as you take exciting steps forward.

So, let's explore how AI can be a powerful tool to open new doors and help you move forward in ways you may never have considered.

Step 1: Be Honest With ChatGPT (And Yourself)

Introduce Yourself Like You're Talking to a Friend

ChatGPT works best when it knows who you are. So don't be afraid to overshare. For example, when I started, I typed something like this:

"Hi, I'm someone who's stuck figuring out my next career chapter. I've been a journalist, I'm a perfectionist, and I've spent a lot of time tiptoeing through ideas that didn't work. I love gardening and embroidery, but they feel too small to build a career around. I need help narrowing down my ideas and finding something I'm genuinely excited about."

See what I did there? I told Chat my strengths, my struggles, and my passions. I didn't try to sound impressive—I just got real. And that's exactly what you should do. Think of it like meeting a brainstorming partner who's eager to help, no judgment attached.

Step 2: Start Broad—Ask for a Wild Brainstorm

Ask ChatGPT to Help You Dream Again

The first step isn't about narrowing down ideas—it's about getting them all out there. Use prompts like:

- What are some creative ways to turn [gardening/embroidery] into a career?
- What are career options for someone who loves [insert passion here]?
- Can you brainstorm 10 ideas that align with [my skills/values/passions]?

For example, when I told ChatGPT about my love for embroidery, it gave me ideas like:

- Starting an Etsy shop for custom embroidery art.
- Teaching beginner embroidery classes online.
- Writing a blog about embroidery techniques and tools.

- Designing embroidery kits for beginners.

With gardening, it suggested:

- Hosting workshops for beginner gardeners.
- Selling seedlings or homegrown herbs at local markets.
- Starting a YouTube channel about gardening in Arizona (because apparently, not everyone knows how to grow plants in the desert).

The key here? Don't judge the ideas. Just let them flow.

Step 3: Narrow Down the Chaos

Sort Ideas by Passion, Not Practicality

Once Chat gives you a list, it's time to sort. But here's the trick: Don't pick the 'smartest' idea. Pick the one that makes your heart do a little happy dance.

Ask yourself:

- Which idea feels exciting, even if it's a little scary?

- Which one would I do for free, just because I love it?

For me, the answer was clear. I've always loved creating beautiful things—whether it's a stitched piece of art or a lush garden. Those ideas lit me up in a way that spreadsheets or generic side hustles never could. *I never thought I could turn those passions into a career until Chat shifted my mental framework and opened my eyes to opportunities that simply were not on my radar.*

Refine Your Ideas With Follow-Up Prompts

Not quite there yet? Let Chat help you dig deeper. Try asking:

- "Can you combine these ideas into something more unique?"
- "Which of these ideas could I start small and scale over time?"
- "What are some low-cost ways to explore [specific idea]?"

When I mentioned wanting to keep things manageable while raising a family, ChatGPT suggested:

- Turning embroidery into a hobby-to-income experiment by selling seasonal pieces (hello, holiday decorations!).
- Creating a gardening blog that doubles as a space to document my greenhouse project.

Suddenly, my passions didn't feel small—they felt *doable*.

Step 4: Let Go of What Others Might Think

Stop Editing Yourself

This is the hard part. We've spent so much time worrying about what others might say— *"She's opening an Etsy shop? Didn't she used to be a journalist?"*—that we forget the only opinion that matters is our own. Let ChatGPT remind you of that. Seriously.

Ask it:

- "How can I stop worrying about whether people will judge my idea?"
- **"Why is it okay to follow a passion, even if it feels small?"**

- "What are examples of people who turned 'small' ideas into successful careers?"

For me, this step was huge. Chat didn't just give me ideas—it gave me permission to let go of my inner critic and *start dreaming again*. In fact, let's find out what Chat has to say about the second prompt I highlighted above.

Question: *"Why is it okay to follow a passion, even if it feels small?"*

Chat's response: (And I couldn't agree more!)
"Following a passion, even if it feels small, is okay—more than okay—because it's not about the size of the dream; it's about the meaning it holds for you. Small passions have the power to bring joy, purpose, and fulfillment to your life in ways that grand ambitions often can't. They allow you to focus on the present moment, create something that feels authentic, and add depth to your everyday experiences.

Chasing smaller passions isn't a sign of settling; it's a way to honor what genuinely lights you up without the pressure of external validation or impossible expectations. It's about embracing what feels right for you, in the life you have now, and recognizing that the value of a dream isn't measured by its scope—it's measured by how much it fulfills you."

I Can't Help but Wonder…What Would Julia Do?

It's fascinating to think about how people reinvented themselves *before* AI and the internet. Many of the well-known figures I mentioned earlier in this book pivoted to second careers in a world where resources were limited—no Google searches, no online courses, no ChatGPT to help brainstorm the next move. They had to rely on gut instinct, trial and error, and sheer determination to figure it out.

Take Julia Child, for example. Back then, if she wanted to master French cooking, she had to move to Paris, enroll in *Le Cordon Bleu*, and spend years perfecting her craft. Imagine if she'd had ChatGPT at her fingertips. Perhaps she would have done things differently. She could have asked, *"Explain the science behind a perfect soufflé,"* or *"How can I break down complex French recipes for American home cooks?"* Maybe she would have used it to brainstorm book titles, refine her teaching approach, or map out a content plan for a modern-day YouTube cooking channel. The point is, she found a way to pivot without any of the technology we have today—so just imagine what's possible *for you* with tools like AI to guide and support your journey.

And for those of you who feel a pang of guilt—like using ChatGPT is somehow *cheating* or taking a shortcut—think about it this way: Every generation has had tools that made reinvention easier. Julia Child had cookbooks and mentors. Modern entrepreneurs have podcasts, online courses, and business coaches. AI is just another tool—one that can help you organize your thoughts, spark ideas, and refine your vision. It doesn't replace *your* creativity; it amplifies it. The real work—pursuing your passions, taking action, and putting yourself out there—is still yours to do. AI simply gives you a smarter way to navigate the process.

Not everyone can move to Paris and train at *Le Cordon Bleu* like Julia did, but AI helps level the playing field. Passion isn't limited by location or resources anymore. Whether it's French cooking or a creative business, *AI gives more people access to knowledge and ideas once reserved for the lucky few.*

Write, Reflect, Restart

Step 1: Introduce Yourself to Chat *(Be sure to pay for the most updated and upgraded ChatGPT. It's affordable and it's completely worth it.)*

Write a quick intro about who you are, what you love, and why you're stuck. Don't hold back—be as honest as you'd be with a friend. Go ahead and write it directly into Chat.

Step 2: Brainstorm Wildly

Ask ChatGPT for a big list of ideas, from practical to completely off-the-wall. Challenge Chat for better, more unique ideas.

Step 3: Pick Your Happy Dance Idea

Choose the idea that makes you smile. That's your cannonball idea. No more toe-dipping for us anymore.

Reflections and Next Steps

Here's the thing: ChatGPT isn't here to make the decision for you—it's here to help you see what's possible. It's the brainstorming partner who never gets tired, the coach who reminds you that you've got this, and the cheerleader who believes your passions are worth pursuing.

So go ahead. Share your messy thoughts, your wild ideas, and your fears. Let AI help you sift through the noise and land on the one thing that feels right. *Because sometimes, the hardest part isn't finding the idea—it's giving yourself permission to chase it.*

Chapter 9

From Idea to Action

So, you've picked your cannonball idea. Yay! Now what? This is where most of us get *stuck, again*. We're excited, but we're also overwhelmed. Where do we start? How do we keep the momentum going? And what if we mess it all up?

Here's the good news: You don't have to figure this out alone. ChatGPT can do more than brainstorm—it can help you create a step-by-step plan, stay on track, and even think in ways you never imagined. The trick is knowing how to ask the right questions, steer the conversation, and let AI guide you toward that *aha* moment.

Step 1: Get ChatGPT on the Same Page

When it comes to working with ChatGPT, the secret sauce lies in how you frame your prompts. You got a taste of that last chapter, but let's go way deeper now. Think of it like giving directions: the clearer and more detailed you are, the more precise and helpful the response will be. If you ask something vague, like "How do I start a business?" you'll get a generic answer. But if you're specific, such as, "I want to start a business selling hand-painted pottery online. Can you help me come up with a name, a target audience, and a marketing plan that keeps costs under $500?" you'll get a tailored, actionable response.

The magic happens when you treat ChatGPT like a collaborative partner—share your passions, your goals, and even your concerns. For example, if you're an overthinker who has abandoned past projects, tell it that! Say, "I tend to overthink and get stuck. Help me keep this plan simple and manageable." This level of transparency helps steer the conversation in the right direction, ensuring the advice you get is aligned with your needs and tailored to your personality. The more you guide ChatGPT with thoughtful, layered prompts, the better it can help you take your ideas to the next level.

Give It Context

Start by reminding ChatGPT where you're at and what you want. This isn't just about repeating your cannonball idea—it's about setting the stage for what comes next. For example, I told ChatGPT:

"Hi again! I've decided to focus on turning my embroidery and gardening passions into a career. I want to create a 30-day plan to explore these ideas and build momentum. I'm looking for low-cost, creative ways to get started while balancing family life and managing my perfectionism. Can you help me?"

Notice how I didn't just say, "Help me make a plan." I gave it the full picture: my goals, my challenges, and what I need help with. This helps ChatGPT tailor its advice to you—and keeps it from going off in the wrong direction.

Ask for Next-Step Ideas

It's time to move beyond just thinking about your idea and start asking ChatGPT for concrete next steps. This is where Chat truly shines because it can take all those swirling thoughts in your head and turn them into a roadmap. For example, try prompts like:

- What are five specific steps I can take to test out my idea for [a baking business, a photography side hustle, writing a children's book, or launching a fitness blog] without spending a lot of money?
- Can you help me brainstorm niche ideas within [my interest] that might appeal to a specific audience?
- What's a realistic 30-day plan to make progress on this idea while balancing a busy schedule?

For instance, imagine you're passionate about baking and want to explore it further. Chat might suggest steps like:

- Perfecting three signature recipes that showcase your unique style.
- Setting up a simple Instagram page to document your creations and gather feedback.
- Researching pop-up events or local farmer's markets to test out small batches.
- Creating a short list of affordable tools or supplies to upgrade your setup.

- Asking for creative feedback on your branding—even suggesting a catchy business name.

The beauty of this process is that Chat isn't here to judge or overcomplicate things. It's here to give you options, structure, and motivation when you feel stuck. And the best part? Once you have those steps, suddenly, it doesn't feel so overwhelming anymore. You're not staring at a blank page or an endless to-do list; you're looking at clear, manageable actions to bring your passion to life.

Step 2: Push ChatGPT to be More Creative

Challenge It to Think Outside the Box

To truly blow the roof off your idea, you need to push ChatGPT to go beyond the obvious. Think of it as your creative collaborator, the one who's not afraid to take your seemingly simple concept and turn it into something extraordinary. Start by asking Chat to help you explore ideas that will make your project unique, engaging, or even groundbreaking. For example, try prompts like:

- What's a completely unexpected way to approach [my passion for photography, jewelry-making, or baking] that no one else is doing?
- How can I combine [two unrelated interests, like knitting and travel, or fitness and cooking] into one standout idea?
- What's a way to generate income from this that feels innovative and fresh?

When I asked ChatGPT how I could elevate a basic embroidery idea, the results blew me away:

- It suggested creating embroidery kits with designs inspired by nature and pairing them with starter seed kits for an interactive "stitch and grow" concept.
- It proposed hosting hybrid workshops—imagine an evening where you learn gardening basics and embroidery techniques all in one go. Picture this: "Stitch and Grow Nights" under twinkling lights in a backyard garden setting.
- Chat even recommended offering custom embroidery pieces that feature personal garden

quotes or loved ones' favorite flowers, turning a hobby into a deeply sentimental (and profitable) service.

These ideas pushed me to see beyond the simple act of gardening and handmade art, inspiring a complete shift in my mindset. They opened doors I didn't even know existed. By challenging ChatGPT to think outside the box, it helped me shift from basic to the truly inspired. You can do the same, whether your passion is photography, fitness, writing, or something completely niche.

The best part? Chat isn't limited by your doubts. While we often dismiss our ideas as "too small" or "already done," ChatGPT sees possibility everywhere. *All you need to do is dare to ask it for more.* Tell it to think bolder, and better—then sit back and watch as your simple seed of an idea blossoms into something incredible.

Step 3: Get Unstuck When ChatGPT Goes Off Track

Steer It Back With Specific Prompts

When ChatGPT veers off course, don't see it as a dead end—see it as an opportunity to fine-tune. Think of Chat as an enthusiastic brainstorming partner who occasionally gets a little too creative (or a little too generic). Your job? Guide it back on track with clear, specific prompts.

For instance, let's say you're exploring the idea of starting a photography business, but Chat keeps steering the conversation toward wedding photography when you're more interested in creating fine art prints. Instead of getting frustrated, try prompts like:

- This doesn't feel quite right. Can you suggest ideas for selling photography as fine art rather than event-based work?
- I'm looking for ideas that require minimal start-up costs but align with my love of landscapes. What would you recommend?

- Can you suggest marketing strategies that don't rely heavily on social media but focus on in-person sales or gallery partnerships?

Redirecting Chat isn't about starting over; it's about refining the process until the ideas feel uniquely yours. Once you master this skill, ChatGPT transforms from a tool into a trusted collaborator—one that helps you uncover fresh perspectives and stay aligned with your passions.

Now, take that leap! Whether your first step is a list of supplies or creating a sketch idea for your new logo, every tiny action builds momentum. *And when Chat wows you with its insights, don't forget to give yourself credit— you're the one turning its brilliance into something extraordinary.*

Your Next Chapter Starts Here

Before you close this chapter, let's put these ideas into action. **Use the following prompts to ask ChatGPT to refine, expand, or simplify your cannonball idea.**

Here's your template: *(Write down your takeaways from Chat.)*

- **Your Idea:** "Starting a candle business with eco-friendly materials…"

- **Prompt to Expand:** "What's a unique spin I could put on this idea to stand out?"

- **Prompt to Simplify:** "Can you suggest a small, manageable first step to get started?"

- **Prompt to Align:** "How can I focus more on creativity and less on profitability in the early stages?"

- **Prompt to Refine:** "This suggestion doesn't quite fit. Can you adjust it to better align with my goals of [fill in the blank]?"

Chapter 10

Making It Real

Now that you are clear about how to prompt and communicate with Chat, let's start building your next chapter career. Did you know Chat can help generate a company name that reflects your personality and goals? Or create a step-by-step business plan that aligns with your time, energy, and resources? It can even design a 30-day action plan to help you build momentum, stay engaged, and keep moving forward—especially on the days when the thought of doing anything feels like too much.

*(I promise, I'm not getting paid to sing ChatGPT's praises—it's just **that** good. Honestly, it feels like a breath of fresh air compared to all those cookie-cutter self-help books that basically say the same thing in slightly different fonts. **Chat actually helps you**, and I'm genuinely excited to share how.)*

Creating Your Business Name

Creating a company name can feel like a massive hurdle, but Chat makes it easy—and fun. Let's say you've decided to start a side hustle offering meal prep services for busy parents. You can ask Chat prompts like:

- Can you suggest creative business names for a meal prep service that emphasizes healthy, family-friendly meals?
- I want a name that feels approachable and fun but still professional. What ideas do you have?

And if you don't love the first round of ideas, no problem. Tell Chat what you do or don't like about them, and it will refine the suggestions until you hit that *aha* moment. A great name isn't just about sounding cool—it's about capturing your mission and standing out in a crowded market. And Chat is a pro at making that happen.

Building Your Business Plan

Once you have your idea and name, it's time to map out your next steps. ChatGPT can craft a business plan tailored to your goals, whether you're starting small or going all in. For example:

- Can you create a simple business plan for a pet photography service?
- I want to offer virtual coaching sessions for beginner writers. Can you outline how I could structure my services and pricing?

Chat will help you identify your target audience, create a pricing structure, and even suggest ways to keep your start-up costs low. It does this by pulling from tons of examples, case studies, and real-world business strategies it's been trained on—then tailoring the advice to match the details you give it. And because it's customizable, you can ask for adjustments. For instance, if a recommendation feels too complicated or expensive, just ask, "How can I simplify this step?" Remember, Chat is giving you a jumping off point for you to take and make yours.

Building a 30-Day Momentum Plan

Now, let's talk about momentum. Chat can create a personalized 30-day plan to help you stay engaged without burning out. Here's how it works:

- **Set the Goal:**
  Ask Chat, "Can you design a 30-day plan to help me launch my [idea]?" Be specific about what you want to achieve by the end of the month—whether it's building a website, testing a product idea, or growing an audience on Instagram.
- **Customize for Your Schedule:**
  Let Chat know how much time you can realistically commit each day. For example:
    - "I only have 30 minutes a day. Can you create a plan that fits into my schedule?"
    - "I can dedicate weekends to this project. Can you structure my plan accordingly?"

Actionable Steps:

Chat will break down the goal into daily tasks. For example, if you're starting a graphic design business:

- Day 1: Create a Canva account and explore templates.
- Day 2: Sketch out your first design idea.
- Day 3: Research platforms for showcasing your portfolio.

Every day, you'll take on just one small task—no pressure, no chaos, just steady progress. I started with ten minutes a day. That's it. But here's the best part: when there's no overwhelm, those ten minutes have a sneaky way of turning into two focused, fulfilling hours.

Tailoring It to Your Life

What makes Chat truly powerful is its ability to adapt to *you*. Tell it about your schedule, your responsibilities, and your struggles. For example:

- "I'm a mom juggling three kids and a part-time job. How can I realistically make time for this?"

- "I tend to overthink things. Can you give me a simpler plan to follow?"
- "My budget is tight. Can you suggest ways to keep my start-up costs as low as possible?"

Chat will respond with ideas and solutions that fit your unique situation. It's not about doing everything perfectly—it's about finding a way forward that works for your life.

Marketing and Branding Strategies

Chat isn't just for planning—it's also a game-changer for marketing your idea. According to Transcend Digital, small start-ups typically allocate around 10% of their planned annual revenue to marketing, which could add up to thousands per month. In contrast, using Chat for marketing can cost as little as $20 to $100 per month, depending on the platform and features. (Yes, that's a massive amount of money you're about to save!)

Ask questions like:

- "What's a creative way to promote my handmade jewelry business on social media?"
- "How can I market a virtual fitness program for busy parents on a small budget?"

Chat can provide you with content ideas, ad copy, and strategies tailored to your audience. It can even help you design a social media schedule, complete with post ideas and hashtags. And if you're not sure where to start, try asking, "What's the best way to promote my [idea] if I've never done marketing before?"

Thoughts to Take With You

So, what's your next step? Write it down! Maybe it's picking a name, designing your first product, or simply asking Chat for a pep talk to keep going. Whatever it is, start today. Because with the right tools and a little momentum, your idea can go from a "someday" dream to something truly amazing. By the way, doesn't it feel good to start dreaming again?

Chapter 11

When You Want to Quit (Again)

(Read this chapter only if your inner quitter is lurking. If you're feeling good, keep riding that wave and skip to the next chapter—you can always pop back here if fear, doubt, or the itch to jump ship sneaks up on you.)

Ah, the quitting phase—welcome back, old friend. You've tried your next chapter idea. You gave it a shot. But it's not living up to your expectations. It's not making money. It's not filling the void left by your old job/life. And honestly? It's starting to feel like a waste of time.

So now you're staring at your half-finished embroidery project, your podcast mic, or your unopened sourdough starter kit, thinking, "Screw it. I'll just try something else." But before you do, let's pause. Because the real question isn't, "Should I quit?"—it's, "Have I given this a fair shot?" And let me tell you, the answer isn't as clear-cut as you think.

The Psychology of Quitting Too Soon

Let's face it: quitting too soon is a pattern many of us know all too well. *It's not because we're lazy or incapable—it's because the weight of everything else in our lives often overshadows our own dreams.* These days, our motivational speeches are almost always directed outward. We're the ones cheering at soccer games, pushing our partners to chase their goals, and reminding our friends how amazing they are. But when it comes to motivating ourselves, we're running on empty. And when you've had a string of abandoned ideas or half-finished projects, that inner quitter can feel louder than ever.

Here's where AI comes in. It's like having a motivational coach in your back pocket—ready to give you a pep talk whenever you need it. And the real magic happens when you're honest with it about why you're feeling stuck or ready to throw in the towel.

ChatGPT won't judge you. Instead, it will give you perspective, encouragement, and practical steps to move forward. Try prompts like:

- "I feel like quitting. Can you remind me why this idea matters?" *This will help you reconnect with your why—the reason you started this journey in the first place.*
- "What's a small win I can celebrate today to keep going?" Sometimes, all we need is to acknowledge the little victories, like finishing a single task or just showing up.
- "Can you give me a pep talk about staying consistent, even when I feel overwhelmed?" Chat can help you remember that big changes happen through small, steady efforts—not all at once.

From Overwhelmed to Unstoppable

I want to share this story. I was almost done writing this book when my inner quitter came lurking. The moment I started thinking about the complexities of self-publishing—something I had never done before—I felt completely overwhelmed. Hiring someone was out of the question—

too expensive, and honestly, I'm too picky. I needed to figure out how to create the cover art, build a landing page, purchase ISBNs, record the audiobook, and upload everything correctly. The moment my perfectionism kicked in, so did the spiral. I thought, *"If I can't do it perfectly, maybe I shouldn't do it at all."* I nearly pulled the plug on the entire project.

That little voice in my head started whispering:

- *"No one's going to read this anyway."*
- *"By the time I finish, there will be a hundred books just like it."*
- *"Someone with a bigger name and better connections will land a real publisher and write a similar book."*

The thoughts were relentless. I didn't even realize I was falling into my old pattern—getting excited, diving in, then talking myself out of it, convinced that moving forward would just be a waste of time. The same pattern I've been repeating since I walked away from my journalism career four years ago.

I told my husband how complicated doing my own artwork would be. I told him I was going to self-publish and that probably no one would read the book. I told him that after all this effort, I'd be lucky to make a few thousand dollars. And then I said, *"What's the point? This is a waste of my time."*

And just like that, I was back in *stuck* territory.

But this time was different. As I heard the words coming out of my mouth—telling my husband all the reasons this book wasn't worth finishing—I caught myself *in real time*. I wasn't just venting. I was watching myself slip right back into *stuck* mode, falling into the same old pattern. But instead of letting it happen, I finally saw it for what it was.

I covered my face with my hands. *Oh my gosh. This is what my book is about.* If I couldn't be perfect, if this wasn't going to be an epic success, if I wasn't about to change the world... then why bother?

I recognized the pattern as it was happening. But it was my husband who pointed out what I couldn't yet see—

that *this* moment was exactly what needed to go in the book. That I had just lived out the very thing I was writing about. That this book wasn't just for other people—it was helping *me*, too.

And that's when I knew: *No. Not this time.*

I wasn't going to toe-dip and then quit. I love this book. I *want* to finish. I want it to help other people—just like it helped me.

So, I did something different. I sat down at my computer and told Chat exactly how I was feeling. I explained that the artwork and publishing process seemed overwhelming, and I asked it for a simple roadmap. Seconds later, I had a step-by-step plan laid out for me.

Suddenly, it wasn't *too much*. It was doable. The plan was clear:

1. Finish the book.
2. Have friends and family read it for feedback.
3. Create the artwork and typography.
4. Upload it.

And just like that – my book would be finished and available to the world.

It's a lot of steps, but it's manageable. And if I go back to my original daily goal of just *ten minutes a day*, I know I can get it done.

Writing has always been my passion, and publishing this book is a dream come true. But what's even more exciting is how writing it helped me redefine what passion looks like in this season of my life. In my twenties, I wanted to change the world. Now, I've learned to embrace my *smaller, doable and more personal* passions—ones that bring me joy without needing to be groundbreaking. And thanks to Chat, I have a tool to help me build a second career around my passions *without letting my perfectionism get in the way.*

In a strange way, I'm *grateful* that I almost quit. Because it showed me that what I've written in this book isn't just a theory—it's real. Recognizing my patterns is what stopped me from slipping back into them. And now, maybe this book can help *you* do the same.

I didn't quit. I'm not stuck anymore. I have a vision, a plan, and this time, I'm following through.

What It Means to Give Your Idea a Fair Shot

I'm all for cheering you on, but we also need to recognize that sometimes walking away *does* make sense. The key is knowing when it's the right move and when you're just giving up too soon.

Before you throw in the towel, take a step back and assess where you really are. Have you given yourself enough time to see results? (Hint: two weeks isn't enough.) Have you made any progress, even if it's small? *And most importantly, do you still care about this idea, or are you just overwhelmed by the process?* These questions can be tough to answer, but they're essential for figuring out if it's time to push forward or pivot.

This is where ChatGPT becomes a valuable tool – again! Use it to reflect on your progress and pinpoint what's holding you back. Ask it to create a realistic timeline—whether that's three months, six months, or whatever feels manageable for you. Chat can also help you track your progress and, perhaps even more importantly, remind you to celebrate those small wins along the way. Sometimes, all you need is a little structure and encouragement to keep going.

But here's the reality: if your idea feels draining, frustrating, or completely disconnected from what matters to you, it might be time to move on. That said, don't let fear or impatience make the decision for you. Quitting this time should be different from quitting in the past. It's not about giving up out of frustration or overwhelm—it's about recognizing when an idea no longer aligns with your goals or values.

Before calling it quits, let ChatGPT help you brainstorm a potential pivot. You can ask things like:

- "This isn't working for me. Can you suggest a new angle to approach it?"
- "What's a related idea that might be a better fit for my skills and interests?"
- "Can you help me decide if this is worth pursuing further?"

Quitting doesn't have to mean giving up—it can mean recalibrating, refocusing, and finding a path that feels right for you. And sometimes, all it takes is a fresh perspective (or a nudge from Chat) to see your next step clearly.

You got this. Let's push past the negative chatter in your head and focus on the passion again.

Before You Go...

Step 1: Write down one reason you feel like quitting your current idea. (Be brutally honest—it's safe here.)

Step 2: Write down one thing ChatGPT could help you with right now to make that reason feel less daunting.

Step 3: Ask Chat: "How can I give this idea a fair shot before deciding to move on?" What was its best advice?

Chapter 12

This Is What Freedom Feels Like

When I left my career as a journalist, I thought the hardest part would be saying goodbye to the job. It wasn't. The hardest part was the quiet that came afterward. The constant questions—"Do you miss it?" "Are you going back?"—weren't just small talk; they were reminders that everyone, including me, still defined me by what I used to do.

It felt like my entire identity was locked inside those Emmys collecting dust on the shelf. To my kids, they were just shiny decorations. They didn't know about the late nights, the sacrifices, or the constant pressure to perform—they only saw *Mom*. And while being "Mom" is the most important title I'll ever hold, I couldn't shake the feeling that *my story had stopped moving forward.* The most exciting chapters of my life felt like they had already been written, and I was *stuck* just retelling old stories instead of creating new ones. I didn't want my kids to see a mom who had left all her dreams in the past—I wanted them to see a mom who dared to dream *again*.

But I didn't know *how* to dream again.

For so long, I had only ever chased *big* dreams— the kind that made a difference, broke stories, and made a real impact. If it wasn't big, it wasn't worth pursuing. That mindset worked when I was building my career, but after I left journalism, dreaming on that scale felt impossible. The weight of daily responsibilities, of raising a family, made the idea of chasing something *huge* overwhelming. And so, without realizing it, I stopped dreaming altogether.

It wasn't until I started writing this book that I began to untangle what had really been keeping me stuck. It wasn't a lack of ambition. It wasn't a lack of talent or drive. It was the idea that *smaller, more personal dreams didn't matter.* That unless I was changing the world, I wasn't doing anything meaningful.

But I was wrong.

Somewhere between the pages of this book, I gave myself permission to see things differently. I let go of the need for my next chapter to be massive, and instead, I focused on something simpler—*does it make me happy?* That single shift changed everything.

Because the truth is, I *am* still dreaming. I'm just dreaming *differently* now.

My passion for storytelling hasn't disappeared—it's just taken on a new form. My creativity isn't gone—it's thriving in my artwork and gardening, in the small, quiet projects that bring me joy. And my ambition? It's still there, but now it's guiding me toward something *sustainable*, something that fits my life instead of consuming it.

For the first time, my dreams—and my drive to build something of my own—aren't driven by the need to prove myself. They don't have to be monumental to matter. And because of that, I finally feel free.

And that's the path forward—not just for me, but for you, too.

The dreams that once inspired you in your twenties, thirties, or at any other stage in your life don't have to be the dreams you carry forever. You're allowed to change. *You're allowed to outgrow the things that once defined you.* And most importantly, you're allowed to dream in a way that fits the person you are *now*.

This isn't about settling. It's about evolving. It's about realizing that sometimes, the most fulfilling dreams aren't always the ones that shake the world—*but the ones that set you free, and make you feel unstuck and unstoppable.*

A Note from Me to You

If this book leaves you with anything, I hope it's a gentler way of seeing yourself—and a renewed belief that your next chapter is worth writing. Feeling stuck doesn't mean something is wrong with you—it just means you're in a moment of transition. And that moment, while hard, can also hold the beginning of something beautiful.

Writing this book helped me see that more clearly than ever. It helped me understand how I got stuck, what I was really craving, and what I needed to let go of in order to move forward. It showed me that dreaming differently isn't giving up—it's growing up. It's learning to honor where we are now, not just where we've been.

Since finishing this book, I've started taking my own small, momentum-building steps—ones I used to talk myself out of. I built the greenhouse I'd dreamed about for years. It didn't happen overnight, and not without challenges—but with intention and heart. I hope to use the space to teach and connect with others through garden-to-table living. I also shared my artwork for the first time—nothing grand, but meaningful to me. A quiet start, and one I almost didn't allow myself to make.

What's been most powerful, though, is hearing from people like you. Readers who are navigating their own pivots, dreaming new dreams, and trusting themselves to begin again. Your stories, your courage, your honesty—they've been a gift.

So, thank you. For reading. For reflecting. For being willing to explore what's next.

I'll keep creating, dreaming, and embracing the unexpected chapters still ahead—and I know you will too.

Because you're not stuck. You're just getting started.

Best wishes,

Abbie

www.ingramcontent.com/pod-product-compliance
Lightning Source LLC
LaVergne TN
LVHW050045090426
835510LV00043B/3095